W9-DJH-886

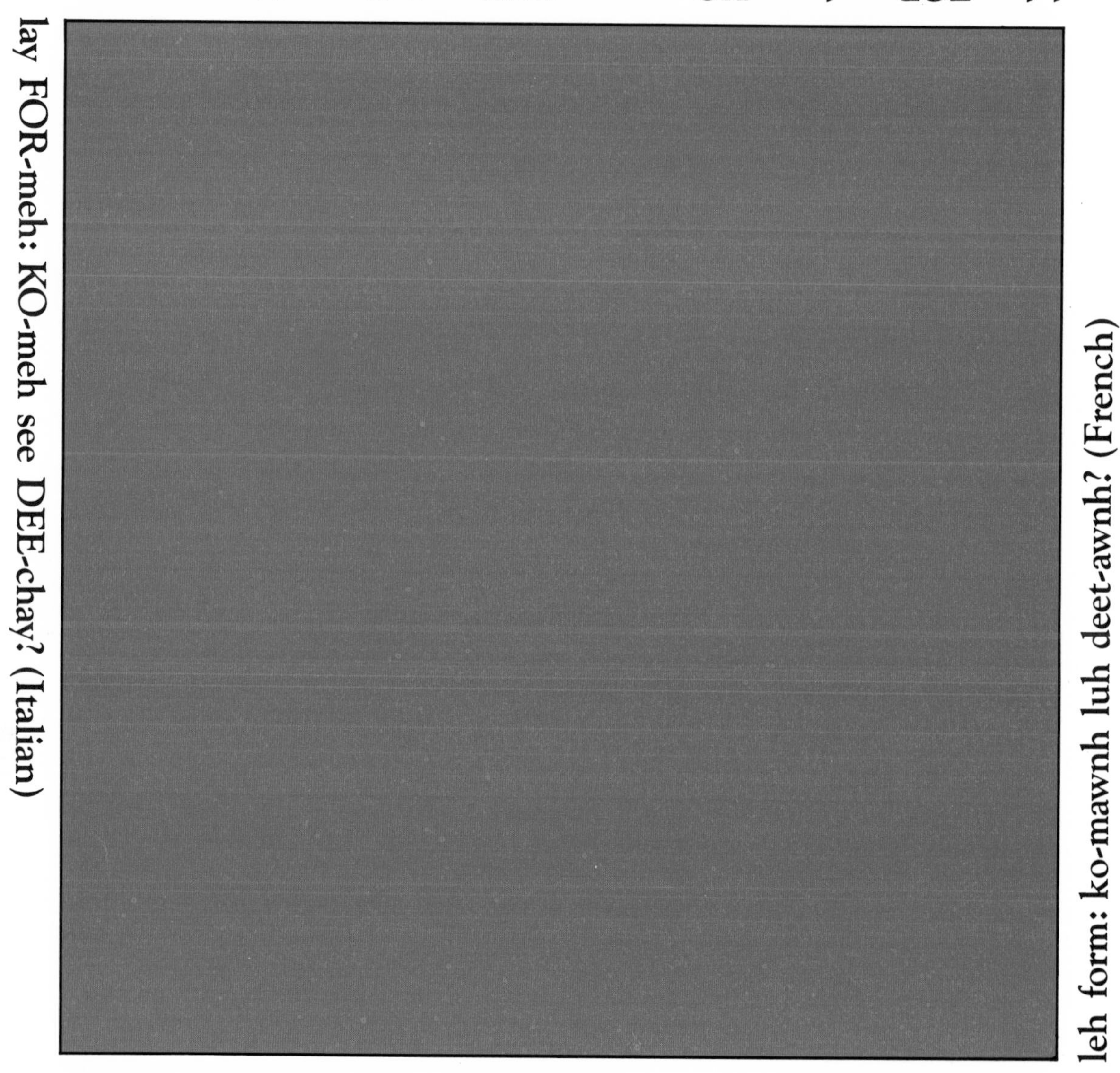

shapes: how do you say it? (English)

leh form: ko-mawnh luh deet-awnh? (French)

lahss FOR-mahss: ¿KO-mo say DEE-say? (Spanish)

lay FOR-meh: KO-meh see DEE-chay? (Italian)

las formas: ¿como se dice?

le forme: come si dice?

# Shapes: How Do You Say It?

English · French · Spanish · Italian

by Meredith Dunham

Lothrop, Lee & Shepard Books New York

les formes: comment le dit-on?

shapes: how do you say it?

Inquiries should be addressed to Lothrop, Lee & Shepard Books, a division of William Morrow & Company, Inc., 105 Madison Avenue, New York, New York 10016. Printed in Japan.

First Edition 1 2 3 4 5 6 7 8 9 10

Library of Congress Cataloging in Publication Data
Dunham, Meredith. Shapes: how do you say it?
English, French, Italian, and Spanish. Summary: Introduces different shapes in illustrations and appropriate descriptive words in English, French, Spanish and Italian.
1. Geometry—Juvenile literature. [1. Shape. 2. Geometry. 3. Picture dictionary, Polyglot]
I. Title. QA447.D86 1987 516.2′2 86-27740
ISBN 0-688-06952-5 ISBN 0-688-06953-3 (lib. bdg.)

To Professor E. Trautvetter in appreciation

circle (English)

luh SEHRkluh (French)

ell SEER-koo-lo (Spanish)

eel CHEER-kee-oh (Italian)

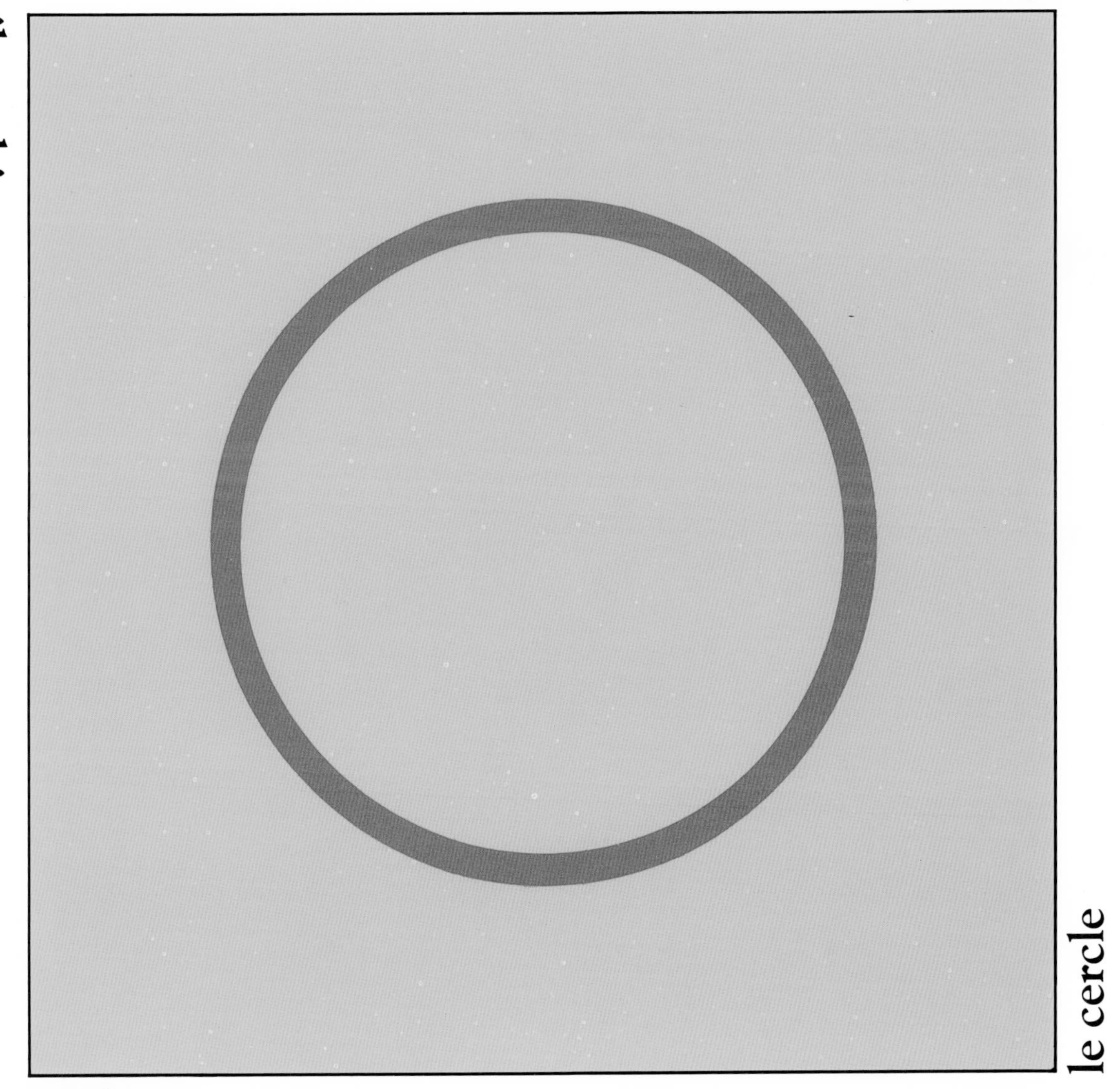

circle

le cercle

el círculo

il cerchio

ell kwah-DRAH-toh (Spanish)

eel kwahd-RAH-toh (Italian)

luh kah-reh (French)

square (English)

el cuadrado

il quadrato

le carré

square

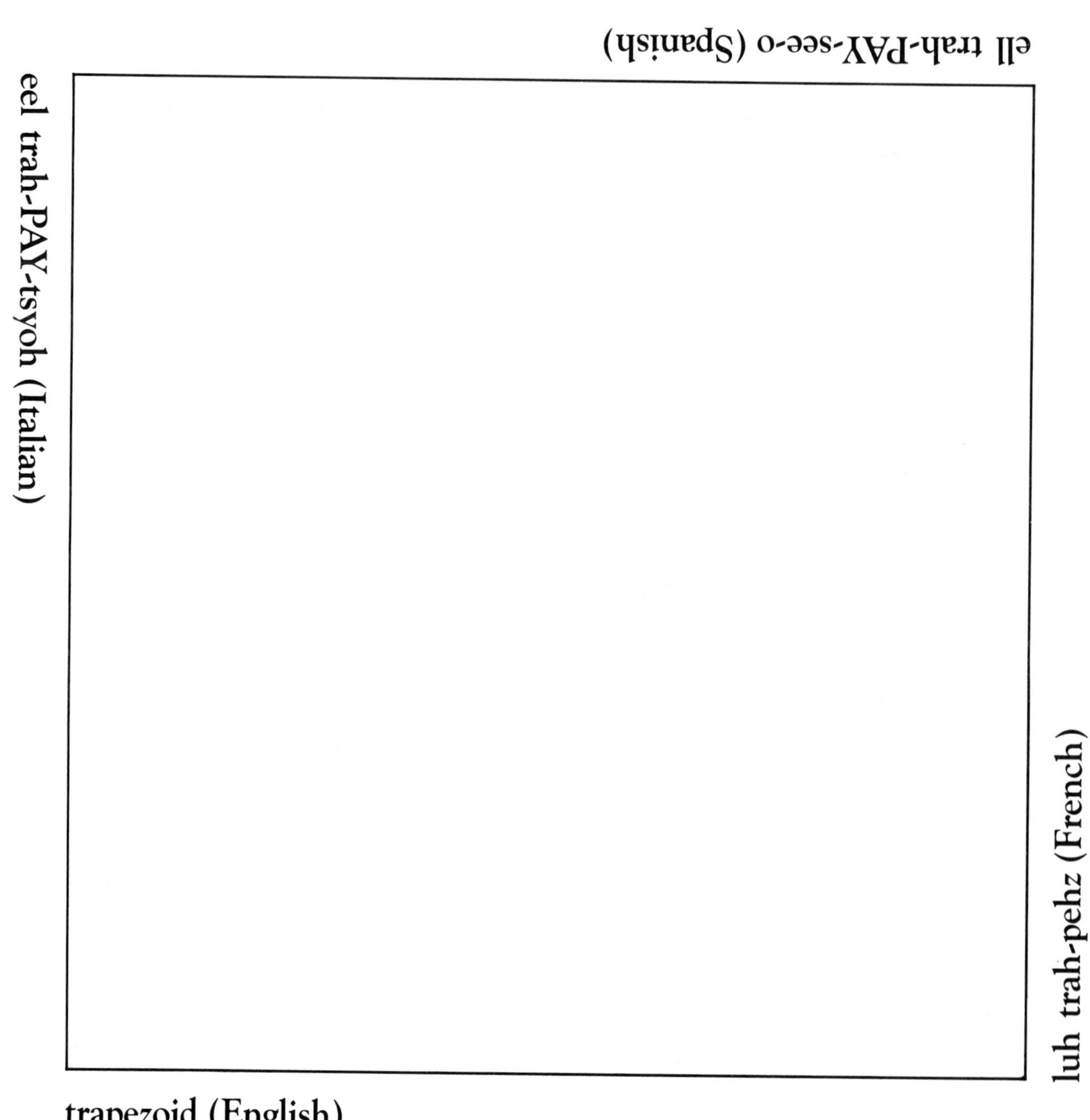


trapezoid (English)

el trapecio

il trapezio

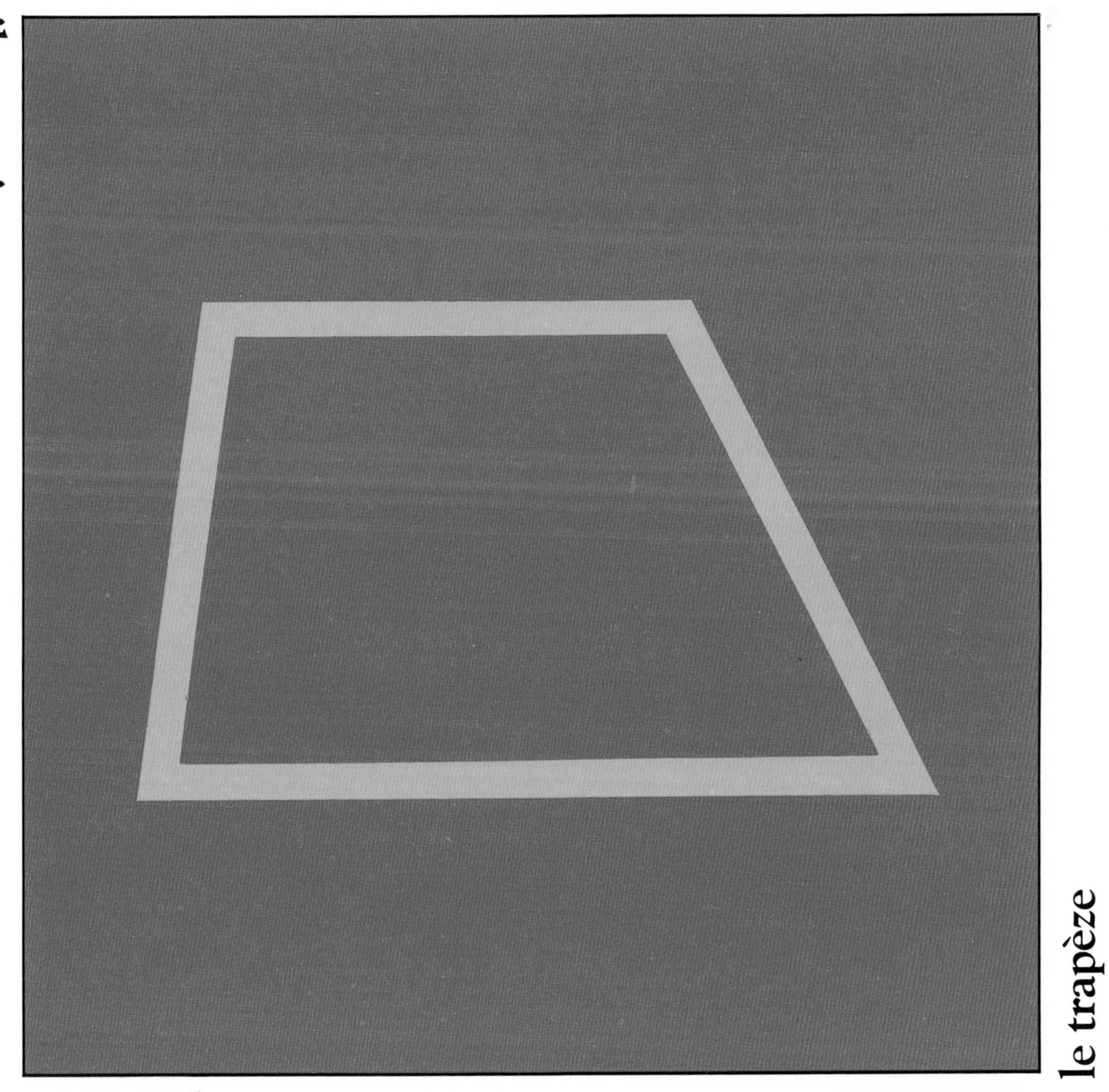

le trapèze

trapezoid

oval (English)

loh-vahl (French)

ell OH-vah-lo (Spanish)

loh-VAH-lay (Italian)

oval

l'ovale

el óvalo

l'ovale

octagon (English)

lok-tah-gun (French)

ell ok-TAH-go-no (Spanish)

loht-TAGH-oh-noh (Italian)

el octágono

l'ottagono

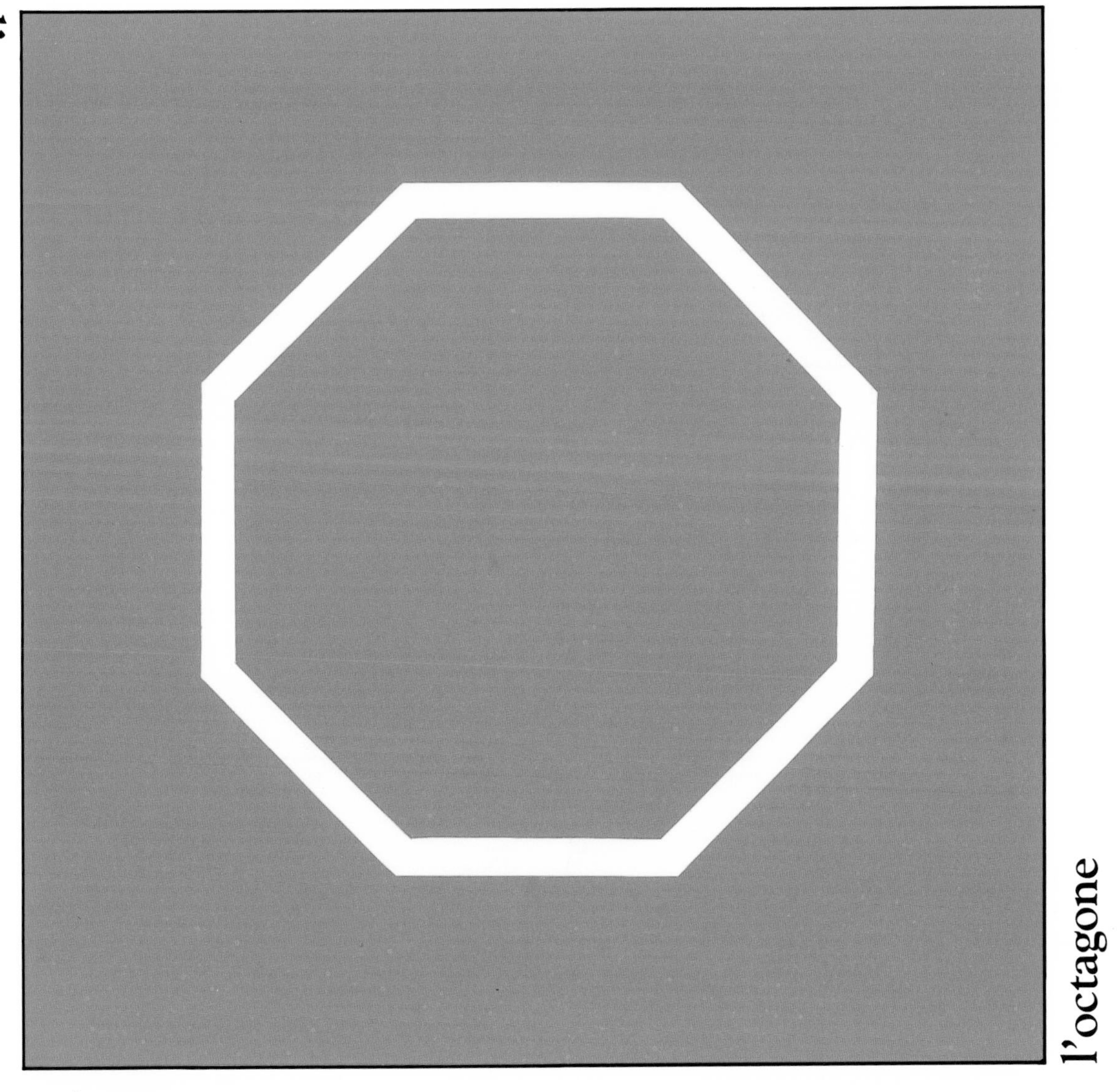

l'octagone

octagon

star (English)

leh-twahl (French)

lah ess-TRAY-ya (Spanish)

lah STEL-lah (Italian)

la estrella

la stella

l’étoile

star

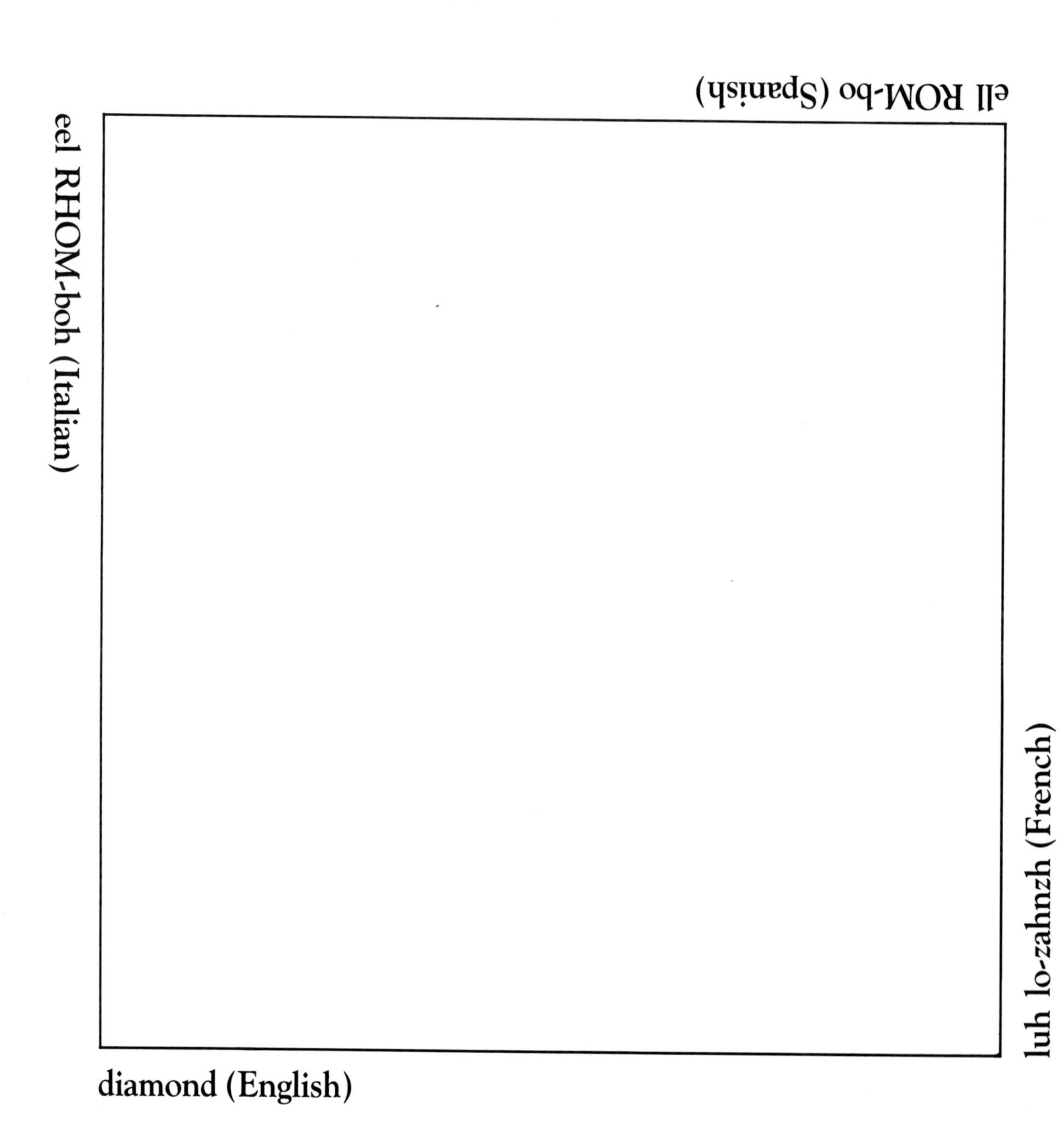
ell ROM-bo (Spanish)
eel RHOM-boh (Italian)
luh lo-zahnzh (French)
diamond (English)

el rombo

il rombo

le losange

diamond

ell tree-AHN-goo-lo (Spanish)

eel tree-ANGH-oh-lo (Italian)

luh TREE-AWNGluh (French)

triangle (English)

el triángulo

il triangolo

le triangle

triangle

heart (English)

luh kuhr (French)

ell ko-rah-SOHN (Spanish)

eel KWAW-ray (Italian)

el corazón

il cuore

le coeur

heart

rectangle (English)

luh REKT-AWNGluh (French)

ell rek-TAHN-goo-lo (Spanish)

eel rayt-TANGH-oh-lo (Italian)

el rectángulo

il rettangolo

le rectangle

rectangle

## A Note on How You Say It

Read the pronunciation guides as if you were reading English text, accenting the syllables in capital letters. This will give the approximate sound of the French, Spanish, and Italian phrases. It can only be approximate because each language has some sounds that do not exist in English.

In French, the R is pronounced far back in the throat. The U (represented here as EW) is pronounced by rounding the lips for OO and saying EE instead. The nasal sound represented as a vowel plus NH (like ANH) is made by saying the vowel "through the nose."

In Spanish, the R is trilled with the tip of the tongue. A double R (RR) is trilled longer than a single R.

In Italian, the R is also trilled. A double consonant is pronounced longer than a single consonant.

Many vowel sounds in English are actually combinations of sounds. For example, if you say the word *make* slowly, you will hear EH and EE in the sound of the *a*. In French, Spanish, and Italian, the vowels are pure—containing only one sound.

If you listen to a native or trained speaker of these languages, you will notice other differences. But that's no reason not to have the fun of saying it in French, Spanish, and Italian!